APPLES OF THE WEL

A list of the old varieties of apples (
orchards of the West Midlands and the 1

Introduction

The main reason for producing this list is to provide guidance on the choice of apple varieties for those wishing to restore an old orchard or to plant a new orchard following the traditional pattern. If you are fortunate enough to own an old apple tree, however decrepit, try your best to keep it alive, at least until it has been identified or a young tree has been propagated from the old one. Your ancient tree may be the sole survivor of a rare breed. Further information is available from the Marcher Apple Network.

Scope

The list contains dessert and culinary varieties; cider apples are excluded unless the variety was also used for another purpose. Only those varieties which are at present available for sale by nurseries are included. However, if the name of an apple is in doubt, that variety is excluded from this list, even if it appears in current nursery catalogues. Extensive research is necessary before an old variety can be authenticated; several are at present under consideration in both national and local collections.

Welsh apples pose greater problems of identification because, with a few exceptions such as Saint Cecilia and Cissy, no detailed descriptions exist and there are no specimen trees in the national collections. There may be a few clues lurking somewhere in the literature, but quite often there is little beyond folk memories, which need to be corroborated by independent witnesses, to establish identity.

Arrangement

In the first part of the list apples are listed alphabetically under the regions (usually counties) in which it is thought they originated. Sometimes more than one county lays claim to a particular variety and in the case of some very old varieties such as Gennet Moyle and Crimson Quoining, their provenance has been lost in the mists of time. Of course, many varieties listed under a particular county have also been cultivated successfully in adjacent regions.

In the second part there is a list of apples which are not, in most cases, local but which have been grown in our area for over a hundred years. These are the varieties we have encountered most often on our visits to orchards in the Welsh Marches - the survivors that have stood the test of time!

Lastly, whatever you choose, make sure that there are other varieties that flower at about the same time planted nearby. Except for a few cultivars still under investigation, pollination numbers given at the end of entries indicate relative flowering times, following the pattern used in ' The Book of Apples'. For example, Golden Spire (*5*) flowers early each spring, whereas King's Acre Bountiful (*23*) is one of the latest to bloom. For effective cross-pollination, trees need neighbours with pollination numbers not more than three digits apart. Triploid cultivars (marked +) need two other varieties with appropriate pollination numbers.

PART 1 - LOCAL APPLES
GLOUCESTERSHIRE APPLES

ASHMEAD'S KERNEL
Medium-sized russet fruit with excellent flavour, at its best between December and February. Raised by Dr. Ashmead of Gloucester about 1700. One of our best dessert apples, with a rich sweet aromatic flavour. It appears to do well even in the wetter parts of the Welsh Marches. *14*

CHAXHILL RED
Primarily recommended for cider, but, like Tom Putt, sometimes valuable as a triple-purpose apple. First recorded in 1873, raised by Mr. Bennett of Chaxhill near Westbury-on-Severn. *17*

HUNT'S DUKE OF GLOUCESTER
Small dessert variety raised by Dr. Fry of Gloucester in early 19th Century. The oblng or conical apples are green ith a brownih-red flush n the sunny side, but the colour is partially obscured by an extensive layer of thin browm russet. With an intense rich flavour, it is best from Christmas until February. *10*

LODGEMORE NONPAREIL
A small, flattened, reinette apple for eating during the early Spring. It was raised in the early 19th Century by Mr. Cook of Lodgemore, near Stroud and sold by the Clissold Nursery as 'Clissold's Seedling'. Although the fruits are very small, they have an excellent flavour. *23*

PUCKRUPP PIPPIN
Neat, medium-sized, conical apple, with a golden skin variably covered with cinnamon russet. It has a rich, intense, 'nonpareil' flavour and keeps in sound condition until January. First recorded in 1872, presumably having been raised or found at Puckrup near Tewkesbury. *8*

SEVERN BANK
Mid-season culinary variety which cooks to an appetizing puree. It is a large, conical, angular apple with a deeply set eye and a smooth yellow skin which can have an orange flush and small red streaks. Described in 1884 and grown extensively in commercial and farm orchards in the Severn Valley in late 19th and early 20th Centuries. *8*

TEWKESBURY BARON
Small, flattened, dark crimson fruit resembling Devonshire Quarrenden, but later ripening. In spite of its imposing name it has only an 'insipid flavour', according to 'The Apple Register'. First recorded in 1883. *7*

WHEELER'S RUSSET
Old dessert variety, probably from the early 18th Century, but confused in the past with Pinner Seedling. The conical angular fruit, often coated with russet, has a sweet, slightly aromatic flavour and remains in good condition until March or April. *12*

HEREFORDSHIRE APPLES

BYFORD WONDER
Mid-season culinary apple, introduced in 1894 by Cranston's (King's Acre) Nurseries, Hereford, which later became Wyevale; large, dark yellow in colour, sweet-sharp, crisp and juicy; keeps shape when cooked, yellowish flesh, flavour sometimes slightly pear-like. Named after a village beside the Wye, a few miles upstream from Hereford. *10*

COLLINGTON BIG BITTERS
Large, roundish, green apple, frequently marked with narrow red streaks. Although principally grown for making a bittersweet cider, it was also used as a culinary variety, and highly prized for making mincemeat. Probably raised in Herefordshire in the 19th Century and extensively planted in the West Midlands. Known around Bromyard as 'The Mincemeat Apple'. *1*

COLWALL QUOINING
Prominently red-flushed, striped and ribbed, this variety was apparently recorded at Tenbury Wells, Worcestershire, in 1949, although its name clearly links it with Herefordshire. It is a mid-season dessert apple which is soft and juicy. *15*

CRIMSON QUOINING synonym HEREFORDSHIRE QUOINING
Suitable for culinary purposes or dessert, this variety, though first recorded in 1831, is thought to have originated in Herefordshire at a much earlier date, where it was traditionally known as 'Herefordshire Queening'. This is characterised as a conical apple, covered with a red flush, sweet, scented and with soft cream flesh - the last apple on the final illustration of 'The Herefordshire Pomona', which affirmed its continuing popularity. The 'Old Quining' illustrated in Knight's 'Pomona Herefordiensis' (1811) appears to be this variety. According to Knight it was cultivated as a cider apple in the 17th Century. *14*

The term 'Quoining' - referring to the quoins of a building - has traditionally been applied to prominently ribbed apples.

DOWNTON PIPPIN
Raised at Elton Hall in Herefordshire in 1792 by Thomas Andrew Knight, pioneer horticulturalist and geneticist, a leading member of the Royal Society (which became the Royal Horticultural Society). His brother Richard Payne Knight built Downton Castle. Downton Pippin is a fairly small, mid-season dessert apple with an intense, sharp flavour. The original tree was planted at Wormsley Grange, childhood home of the brothers. *12*

FORESTER
Late-season cooker with medium to large, flat or conical fruits with indistinct ribs; skin deep green at first, becoming yellow overlain with brown russet. First described in 1883 and extensively planted in the West Midlands in the late 19th Century. *16*

GENNET MOYLE
The name means 'hybrid scion'. This variety was extensively cultivated in the 15th Century, particularly in the Archenfield region, and was generally used for cider making until supplanted in popularity by the celebrated Redstreak, raised by Lord Scudamore in the 17th Century. It continued to be favoured for culinary purposes by those with 'delicate stomachs'! It is a large, rather flattened, golden apple with red stripes. *5 +*

GOLDEN HARVEY
Considered most likely to have originated in Herefordshire early in the 16th Century and familiar to prominent writers in the reign of King Charles II. Small, golden and russeted, with an intense flavour: popular for the Victorian dessert, but eventually considered too small: also called the 'Brandy Apple', since the high specific gravity of the juice made for strong cider. *13*

HEREFORDSHIRE BEEFING
Known in Herefordshire in the late 18th Century - 'beefing' is a corruption of 'beau fin'. As with the Norfolk Beefing, these apples were mainly used for slow baking in a bread oven after the bread had been removed; an iron plate being placed on top to press the air out. In the old days, particularly in Norwich, 'biffins' thus baked were packed in boxes and sent to the London fruiterers for sale as presents. In 2000, the writer of this note received a present of a box of biffins from the owner of a Norfolk Beefing in Cheshire, which though of unpromising appearance, proved to have an exciting flavour. The Herefordshire Beefing is a flattish, dark red, rather dry, late keeper. *17*

KING'S ACRE BOUNTIFUL
Introduced by King's Acre Nursery (see Byford Wonder) in 1904, this mid-season culinary apple is large and round and becomes almost white when ripe. It has a sub-acid flavour and cooks to a puree. Late flowering, disease resistant, and a regular, heavy cropper. *23*

KING'S ACRE PIPPIN
This high-quality dessert apple was raised at King's Acre Nursery and introduced in 1899. It was described in their catalogue as a cross between Ribston Pippin and Sturmer Pippin. The large, roundish, somewhat angular fruit often bears patches of russet and a brownish-red flush. It has an excellent flavour and remains sound until March. *12 +*

LORD HINDLIP
Although named after the owner of a Worcestershire estate, this richly-flavoured, late-keeping dessert apple was first exhibited in 1896 by Mr. Watkins of Pomona Farm, Hereford. A fairly large, handsome, conical apple, speckled and dotted with crimson-red. *10*

NEW GERMAN
Recorded in Herefordshire in 1884. Large culinary apple with deep red flush and sharp flavour; cooks to a puree; prone to bitter pit and not a late keeper. *12*

PIG'S NOSE PIPPIN
Considered to have originated in Herefordshire and described in 1884. Small, with wide and shallow basin, making the top of the fruit flat, like a pig's snout: sweet, crisp and lightly aromatic, late-Autumn dessert apple. *14*

POMEROY OF HEREFORDSHIRE
"The King's Apple" exists in other forms in other parts of the country and is generally considered to be of great antiquity. The Herefordshire version is an early-to-mid season dessert apple with rather soft flesh but an excellent flavour, attractively coloured with light-green, red and russet patches. Known as 'Sugar Apple' in parts of Herefordshire. *13*

SAM'S CRAB
NOT to be confused with crab apples! Used both for cider and cooking in the past, but renowned for its flavour as an early dessert apple. "a prime favourite with all Herefordshire schoolchildren (no mean judges of a good apple)", said 'The Herefordshire Pomona' in the 1880s and "it requires a warm soil and sunny situation to bring it to perfection". Recorded among the London Horticultural Society's fruit collection in the 1830s. *15*

STOKE EDITH PIPPIN
Recorded in 1872 and presumably raised by the Head Gardener of the Foley estate at Stoke Edith. A high quality, mid-late season dessert apple, yellow with an occasional orange flush, sweet and slightly scented. *11*

TEN COMMANDMENTS
Originated in Herefordshire and first exhibited in 1883. Takes its name from the ten red spots which are seen around the core when the apple is sliced transversely. Small, dark-red apples used mainly for cider but with a passable flavour. *13*

TILLINGTON COURT
Raised on Burghill Fruit Farm probably in the early 20th Century: a brightly coloured, prominently ribbed cooking apple with a sub-acid flavour. *15*

TYLER'S KERNEL
Brought to the Herefordshire Fruit Shows by a Mr. Tyler and certificated by the R.H.S. in 1883. Claimed by 'The Herefordshire Pomona' to have been an unrecorded denizen of Herefordshire orchards for a long period, this is a handsome culinary apple, brilliantly flushed and streaked with red, sometimes eaten as dessert in late Autumn. *10*

YELLOW INGESTRIE
Raised by Thomas Andrew Knight (see Downton Pippin) in 1800 and named after Lord Talbot's estate, Ingestrie Hall, in Staffordshire. This is a small, yellow, cylindrically-shaped early dessert apple. It was known in the past as "A lawn tree apple on account of the beautiful drooping habit". Its fruit was used for highly decorative table displays. *9*

SHROPSHIRE APPLES

BROOKES'S
Slow-growing trees which bear regular crops of small, conical, red-striped and russeted dessert apples. Though rather dry, they have a rich, aromatic flavour and keep until the Spring. Recorded in 1820. *11*

ONIBURY PIPPIN
Described in a report of the Great Apple and Pear Exhibition of 1883 as "a perfect model for a dessert apple from its handsome neat looks, golden colour and lasting properties", this attractive apple was raised by Thomas Andrew Knight early in the 19th Century. The pippin is named after the village in Shropshire where one of his nurseries was situated. The apples are roundish in shape and may have a brownish flush and often some fine netted russet. The flesh is soft and creamy-white with a pleasant flavour in October and November. *14*

WORCESTERSHIRE APPLES

BETTY GEESON
Very late, medium-sized cooking apple, flat in shape and slightly ribbed around the open eye. The yellow skin sometimes has a red flush. Stated in 'The Fruit Manual' to have been introduced by Dr. Davies of Pershore in 1854, and grown extensively in commercial orchards in the Midlands during the 19th Century. However, in 'The Herefordshire Pomona' is the comment "said to have been raised from a pip by Betty Geeson, near Belvoir". *11*

CHATLEY KERNEL
Medium-sized, rather flattened, long-keeping apple, its green skin having a brownish-red flush and more-or-less thickly freckled with large russet lenticels. First exhibited in 1894. Taylor refers to it as an eating apple but "devoid of real dessert flavour" and "never worth planting". Bunyard on the other hand considered it a culinary variety but "Hardly worth cultivating". *20*

EDWARD VII
First-rate, large, roundish cooking apple, green at picking time in mid-October but ripening to pale yellow, often with a brownish-red flush. It keeps in good condition until March or April and cooks to a fine-flavoured puree. Raised from a cross between Blenheim Orange and Golden Noble and introduced by Messrs. Rowe of Barbourne Nurseries, Worcester, about 1908. Late flowering, with beautiful pink blossom. *21*

GLADSTONE
Early, round or conical, dessert apple, which is flushed and striped dark red. It is a soft, deliciously-flavoured apple, at its best in late July or early August, and a favourite with birds and wasps. Originally called Jackson's Seedling, after Mr Jackson who found it near Kidderminster about 1780; renamed Gladstone in 1883 after certification by R.H.S. *13*

GREEN PURNELL
Large, flattish, angular, green,dessert apple, which sometimes has a dull orange flush and faint stripes. In season from November until February. *12*

HOPE COTTAGE SEEDLING
Raised by Mrs Oakely at Hope Cottage, Rochford, near Tenbury Wells, in 1900, from a pip of Princess Pippin. A mid-season dessert variety similar to its female parent. Princess Pippin is one of the many synonyms of King of the Pippins. *9*

KING COFFEE
Large, rather flat, dessert variety with a dark red flush. When ripe, in November and December, the soft, coarse-fleshed, apples have a "hint of coffee flavour" according to 'The Book of Apples'. *17*

MADRESFIELD COURT
Large, tall, somewhat angular, red dessert apple, with a pleasant, slightly aromatic, flavour. Ripe from October until December. Probably raised by William Crump, Head Gardener at Madresfield Court near Great Malvern and introduced by J.Carless of Worcester in 1915. *11*

MAY QUEEN
A dessert variety with medium-sized, flat, shiny red apples, borne abundantly on a small, compact tree. The firm, crisp apples have an unusual nutty flavour and keep well into the Spring, Raised by Mr. Haywood of Worcester in 1888. *12*

NEWLAND SACK
This dual-purpose variety is reputed to have arisen, in the late 18th Century, from a pip in pomace discarded from a cider-mill at Newland Court near Great Malvern. The medium-sized apples are rather irregular in shape, with light-brown russet markings and an orange-red flush on the sunny side. They will keep in good condition until well into the Spring and are sweet enough for dessert after Christmas. *10*

PITMASTON PINE APPLE
Probably raised at Witley about 1780 by Mr. White, steward to Lord Foley, but marketed by John Williams of Pitmaston, near Worcester. A small, oblong-conical, golden-yellow dessert apple with fine cinnamon russet and concentric circles of russet marks around its closed eye. In spite of its small size it was, deservedly, extolled by Edward Bunyard for its "..most deliciously scented and honeyed flavour". *12*

Above: Puckrupp Pippin

Below: Wheeler's Russet

Above: Sam's Crab

Below: Pomeroy of Herefordshire

Above: Downton Pippin

Below: King's Acre Bountiful

Above: Onibury Pippin

Below: Brookes's

Above: Madresfield Court

Below: William Crump

Above: Rushock Pearmain

Below: Landore

Above: Pig Aderyn

Below: Marged Nicolas

Above: Lord Grosvenor

Below: Scotch Bridget

WORCESTERSHIRE APPLES (continued)

PITMASTON RUSSET NONPAREIL
A medium-sized, flat, heavily-russeted dessert apple. It has a rich flavour and is at its best around Christmas time. Raised by John Williams of Pitmaston about 1815. *11*

RUSHOCK PEARMAIN
There has been some confusion over its name. In the Brogdale Collections King Charles Pearmain appears to be identical with Rushock Pearmain. The latter was raised at Rushock by the blacksmith, Charles Taylor, about 1820 and the apple was often called Charles' Pearmain in Victorian times. At some later date the apple appears to have been elevated to King Charles Pearmain. It is a handsome, regular, conical apple, with a large, open eye and golden skin nearly covered with russet. It has a distinctive nutty flavour, at its best from November to January. *17*

SANDLIN DUCHESS
Brightly coloured, late, dessert apple, like an improved version of Newton Wonder, according to William Crump (see Madresfield Court) who introduced the variety. It is in season from November until February or March. Raised by Mr. H. Gabb at Sandlin, near Malvern, about 1880. *10*

WILLIAM CRUMP
The result of a cross between Cox's Orange Pippin and Worcester Pearmain is a handsome, medium-large, conical-shaped, dessert apple, with dark red flush and stripes. It is crisp and juicy, with a sharp, aromatic flavour and keeps in good condition until about February. Unfortunately, the variety seems to be prone to scab in the wetter western regions. There is some mystery over its place of origin, as both Madresfield Court (where William Crump was Head Gardener) and Messrs. Rowe of Worcester claimed to be the raisers, but in either event, it is clearly a Worcestershire apple. *11*

WORCESTER PEARMAIN
Well-known dessert apple of regular conical shape, its smooth skin more-or-less covered with bright red flush and stripes. Juicy, sweet and delicious, if left to ripen fully on the tree. Reputed to be a seedling of Devonshire Quarrenden raised by Mr Hale at Swan Pool and introduced about 1873 by Richard Smith of Worcester. In the 20th Century it became very popular and was extensively cultivated to supply the market for early eating apples, but often the flavour and texture suffered from the fruit being picked prematurely. *11*

WELSH APPLES

BAKER'S DELICIOUS
Attractive, conical dessert apple with orange flush and carmine stripes. Ripe in September and possessing a good aromatic flavour, much appreciated by birds and grey squirrels. Reputed to have been found in Wales; it was introduced by Bakers of Codsall near Wolverhampton in 1932. 7

CISSY synonym MONMOUTHSHIRE BEAUTY
Medium-sized, highly-coloured dessert apple, round and bright red, with dark crimson streaks. It has a rich flavour in September and early October but does not keep for long. Raised, about 1800, by Mr. Tampling of Malpas, near Newport, Monmouthshire, who had a sister called Cissy. The apple was first known as Tampling; Monmouthshire Beauty was a later synonym. Bunyard perhaps tasted the apple when it was past its best for he wrote that it was ripe in November and had a "very poor flavour". 10

LANDORE synonym MONMOUTH GREEN
Medium-sized, roundish or slightly angular apples which are green when picked but ripen to yellow, sometimes with a brownish flush. Trees appear to be hardy and remarkably resistant to disease. The dual-purpose apples will keep until after Christmas and are pleasant to eat from November. Landore is mentioned in the diary of Francis Kilvert, Curate of Clyro in Radnorshire, in 1872, and has been grown in the area around the Black Mountains since Victorian times. The variety could be far older. It is usually known as Landore in Herefordshire but Monmouth Green in Breconshire. 15

MARGED NICOLAS synonym MORGAN NICOLAS
This rather angular flat-conical apple has a yellow skin with conspicuous brownish lenticels and often some netted russet. It is a dual purpose apple, ripe from the end of October and keeping until early Spring. The trees with their characteristic fountain-like habit are found in several farm orchards in the Dinefwr region of Carmarthenshire, but the variety does not appear to have been listed by any of the larger nurseries. 15

PIG ADERYN
A roundish or somewhat flattened, red-streaked, early eating apple, similar to Tom Putt in appearance. In a proportion of the apples the stalk projects horizontally from a fleshy bump on the base, so that the apples, when inverted, have a fanciful resemblance to a bird, hence the name 'Bird's bill'. Ripe in September - October.

PIG YR WYDD
This cooking apple is also found in farm orchards in the Dinefwr area. It is medium-large in size and conical, bluntly ribbed and rather angular in shape. The skin is pale green when ripe, with a slight pink flush and broken red streaks on the sunny side. The stout stalk is often deflected sideways by a fleshy swelling on one side of the cavity. In season from October until December. 9

SAINT CECILIA
At picking time in mid-October this regular conical dessert apple is dark green streaked with brownish-red, but its colour changes to yellow with red stripes as the fruit ripens. There is often rough brown russet around the base. It has an excellent aromatic flavour from Christmas until Spring. Raised about 1900 at the nurseries of John Basham and Sons at Bassaleg, Monmouthshire, from a Cox's Orange Pippin seedling. *7*

PART 2 - APPLES ASSOCIATED WITH THE MARCHER APPLE NETWORK REGION

Apples which, though not in most cases originating in our area, have been extensively grown here in the past and are still to be found in local farm orchards.

ADAMS'S PEARMAIN
Although derived, apparently, from a source in Norfolk when recorded in 1826, this variety was considered by eminent authorities of the period to have originated in Herefordshire, where it had been known as 'Hanging Pearmain'. It is a late-keeping, conical, reinette with a flavour described as rich, aromatic and nutty, "considered essential for Victorian and Edwardian desserts". *9*

BEAUTY OF BATH
An attractive, red-striped, dessert apple which is ripe in August, raised at Batheaston, near Bath, and introduced about 1864 by nurseryman George Cooling. In spite of its short season and the tendency for the apples to drop prematurely, it was a popular orchard variety in the early part of the 20th Century. Growers used to pile straw on the ground under the trees to reduce bruising of the fruit. *9*

BLENHEIM ORANGE
Originating from the very edge of the Blenheim Palace estate in Oxfordshire in 1740, this has been an enormously popular dessert apple which can also be cooked to a puree suitable for apple charlotte. Makes a large tree and its hard wood was once used for railway cogwheels. The fruit is noted for its orange flush and distinctive nutty flavour, perfect for the Christmas table. *12 +*

BRAMLEY'S SEEDLING
This very popular cooker was raised, about 1810, by Miss Mary Anne Brailsford of Southwell, Nottinghamshire, who planted the tree in her cottage garden. Many years later, when the cottage was owned by a Mr Bramley, the apple was "discovered" by Henry Merryweather, whose nursery introduced it as Bramley's Seedling. The variety is triploid and makes a large spreading tree, which needs to be given plenty of space. Although 'Bramleys' are available in supermarkets at most times of the year, this variety may be a good choice for less favourable sites; it is such a vigorous variety that it will thrive in places where many other sorts fail. *12 +*

CATSHEAD
Certainly in existence in the 17th Century, this seems to be the nearest surviving relative of the medieval costard (hence costermonger). Its upright, ribbed, - at times square - uncompromising profile causes it to live up to its name and made it handy for packing "together with that slice of pickled pork or bacon" for the original 'ploughman's lunch'. In its classic form it is a green apple with a modest brown blush which can be eaten as dessert near the end of its season in January. Widely grown in cottage gardens. *11*

CHARLES ROSS
Raised by Charles Ross about 1890 from a cross between Peasgood's Nonsuch and Cox's Orange Pippin. Originally Ross named it 'Thomas Andrew Knight' but it was later renamed at the request of his employer. It is a really handsome, shapely, dual-purpose apple almost as large as Peasgood's Nonsuch but with the colouring of its other parent. During October and November the apple has a good, slightly aromatic flavour, but thereafter taste and texture rapidly deteriorate. *11*

COURT PENDU PLAT
This beautifully-flavoured very late apple is reputed to have originated in Roman times and was evidently grown all over France by the 16th Century. One of its many synonyms is 'Garnons Pippin', since it was introduced to Herefordshire by the Sir John Cotterell of the day. It needs to be put on a vigorous root-stock because it is a sluggish grower, but regarded as the "wise apple" since it flowers so late that it avoids most Spring frosts. *26*

DUMELOW'S SEEDLING synonym WELLINGTON
Probably raised in the late 18th Century by Richard Dummeller, alias Dumelow, at Shakerstone in Leicestershire. The large, flat-round, regular, pale-green cooker has a conspicuous open eye and some fruits show an orange flush and a few broken red stripes. The flesh is firm and sharply acidic and the apples keep in good condition until March or April; the variety is often recommended for making mincemeat. Renamed in honour of the 'Iron Duke' sometime after the Battle of Waterloo. *15*

GASCOYNE'S SCARLET
The large, slightly ribbed apples have scarlet and carmine patches contrasting with milky-white background and overlain with a fine waxy bloom. Its beautiful colouring made this a favourite exhibition variety. Old trees still survive in several local farm orchards, but they bear only light crops. It is reputed to be more productive on chalky soils. The fruit keeps until December and has a sweet, slightly balsamic flavour. According to Edward Bunyard, the variety was raised by Mr Gascoyne of Bapchild Court, Sittingbourne, and introduced by Messrs G. Bunyard and Co., Maidstone, in 1871. *14*

GOLDEN SPIRE synonym TOM MATTHEWS
Widespread in farm orchards in our region, this tall, bright yellow apple can be used both for cooking and making cider. The variety was valued because it crops regularly even when growing under adverse conditions. Its place of origin is uncertain but probably it was found in Lancashire in mid-19th Century and introduced by the nurseryman Richard Smith of Worcester. In our area it was planted as a cider apple under its synonym 'Tom Matthews'. Like Yellow Ingestrie, it was recommended in late Victorian times, as a decorative shrubbery tree. *5*

GRENADIER
Possibly the best of the 'Codlins' and still quite widely grown in commercial orchards, this was described as "A promising new apple" after George Bunyard had shown it at the Exhibition in 1883. The large fruit, which are somewhat flattened and irregular in shape, turn from pale green to yellow as they ripen. The largest apples can be picked in late July but the fruit will hang on the tree and remain usable until well into September. *11*

KESWICK CODLIN
Originating in Lancashire in the late 18th Century, this favourite pale yellow early cooking apple has been grown extensively in farm orchards in the area: highly decorative by virtue of early blossom and "neat habit". In former times cooked "hot codlins" were sold hot on the streets of London. Described as "...the best early in Herefordshire as an orchard tree" in a Woolhope Club report in 1883. *10*

KING OF THE PIPPINS synonym GOLDEN WINTER PEARMAIN
Another favourite farm-orchard variety used both for dessert and cider-making. When ripe in late Autumn the apples range from round to pearmain-shaped, with skins coloured golden, shaded with reddish-brown and crimson streaks. The flesh is crisp and juicy, with a nutty, slightly bittersweet, flavour. There is some doubt over its provenance. The variety has many alternative names: in Herefordshire it is often known as Prince's Pippin; Shropshire Pippin is another synonym. Some authorities claim that there are two varieties at large which have become thoroughly muddled. *13*

LADY SUDELEY
This very colourful early dessert apple is conical in shape, with a yellow skin boldly splashed and striped with bright scarlet. The flavour is sweet and slightly aromatic in late August and early September, but soon fades. It was raised about 1850, by Mr Jacobs of Petworth, Sussex, and at first called 'Jacob's Strawberry'. It was renamed Lady Sudeley when George Bunyard started to market the variety in 1885. *13*

LANE'S PRINCE ALBERT
This late-keeping cooking apple was introduced by Messrs. Lane of Berkhampstead about 1850. It is notable for its very smooth yellowish-green skin, sometimes coloured with narrow broken orange-red stripes. The flesh is very juicy and acidic and does not break down to a smooth puree when cooked. *12*

LORD DERBY
This large conical, often flat-sided cooker, which can closely resemble Catshead, was first recorded in 1862. The apples turn from bright green to yellow as they ripen. It is a useful cooker to follow Grenadier, in season from October to December. *14*

LORD GROSVENOR
Noted as "new" in John Scott's catalogue of 1873, this large, very angular, pale-green codlin ripens to a very pale cream. It is an excellent cooker and in season for a long period - from August to November - but the apples become very greasy in store. Like most codlins, it bears heavy crops from an early age. *10*

MERE DE MENAGE
Known in the late 17th Century as the 'Flanders Pippin', indicating that it came from the Low Countries. Later called Lord Combermere, after one of Wellington's generals, who presumably imported the variety. It is a large, rather flattened, purplish-red culinary apple, keeping until January: quite widely grown, mostly in gardens. *10*

MONARCH
A popular culinary variety which was raised in 1888 by Messrs Seabrook of Chelmsford, Essex. The trees are vigorous and seem to thrive even in the wetter, western parts of our area. The apples are large, with a prominent open eye, and a smooth skin with pink flush and some broken dark-red stripes. They cook to a well-flavoured puree but need less sugar than 'Bramleys' a quality which contributed to the popularity of Monarch as a cottage-garden tree in the early 20th Century. *11*

NEWTON WONDER
This large, regular, cooking apple is often brightly coloured with a red flush and stripes. The open eye is distinctive and 'king' fruits usually have a characteristic fleshy swelling at the base of the stalk. The apples keep until about March, becoming very greasy in store, but are good to eat in the Spring. The variety originated, about 1870, as a seedling growing in the thatched roof of an inn at King's Newton, Melbourne, Derbyshire. *10*

SCOTCH BRIDGET
As its name implies, this hardy variety seems to be of Scottish provenance, but was widely grown in the North of England and West Midlands in the 19th Century. The large conical apples, which are strongly ribbed and knobbed, develop a bright red colour on the sunny side. It is a useful dual-purpose fruit, in season from October to February. *11*

TOM PUTT
Named after the Rector of Trent in Somerset about 1790, but popular throughout our area in former times, this is a hardy variety and old trees survive in many farm-orchards, even in wet and wind-swept situations. The apples are ribbed and irregular in shape, with bold red stripes; a cottage-garden apple par excellence since it may be eaten, cooked, or made into cider. *10*

RIBSTON PIPPIN
This high-quality eating apple was raised at Ribston Hall in Yorkshire from a pip brought back from Rouen about 1700. The apple is rather irregular in shape, ribbed and flat-sided, but attractively coloured with orange flush, broken red stripes and patches of russet. It is ripe from October until Christmas and has a delicious aromatic flavour - sharper and juicier in the early part of its season. *8 +*

WARNER'S KING
This culinary variety, known since the 18th Century, probably originated in Kent. The very large, irregular, flattened apples, mainly green but sometimes with a brownish-red flush, will keep until January. The tree has large leaves and shows vigorous growth In former times it was a favourite cottage garden variety but is rather prone to canker, especially in the west of our region. *7 +*

REFERENCES

Pomona Herefordiensis - T.A.Knight 1811

The Orchardist - J. Scott 1873

The Herefordshire Pomona - H.G.Bull & R. Hogg published 1876-1885 by the Woolhope Club

The Fruit Manual (5th Edition) - R. Hogg 1884

Journal of the Royal Horticultural Society

Transactions of the Woolhope Naturalists' Field Club

A Handbook of Hardy Fruits - E.A. Bunyard 1920

The Apples of England (3rd Edition) - H.V. Taylor 1946

National Apple Register of the United Kingdom - M.W.G. Smith 1971

The Book of Apples - J. Morgan & A. Richards 1993

A list and descriptions of apples planted at Berrington Hall (National Trust) by S.F.Baldock for the National Council for the Conservation of Plants and Gardens c.1990

Marcher Apple Network

The Marcher Apple Network, a company limited by guarantee, was established in 1993. Through various initiatives, MAN promotes the conservation of the wide range of fruit varieties formerly grown in the traditional orchards of the Welsh Marches.

For further MAN information, benefits of joining MAN, other MAN publications and reference materials (including the Herefordshire Pomona reproduced on CD-ROM) plus lots of apple facts and other contacts please........

.... look on the Marcher Apple Network website:

www.marcherapple.net